Your Guide to Success in Math

Complete Step 0 as soon as you begin your math course.

STEP 0: Plan Your Semester

☐ Register for the online part of the course (if there is one) as soon as possible.

☐ Fill in your Course and Contact information on this pull-out card.

☐ Write important dates from your syllabus on the Semester Organizer section on this pull-out card.

Follow Steps 1–3 during your course. Your instructor will tell you which resources to use—and when—in the textbook or eText, *MyMathGuide* workbook, videos, and MyLab Math. Use these resources for extra help and practice.

STEP 1: PREPARE: Studying the Concepts

☐ Read the textbook or eText, listen to your instructor's lecture, and/or watch the section videos. You can work in *MyMathGuide* as you do this, saving all your work to review throughout the course.

☐ Work the Skill to Review exercises and/or watch the videos in MyLab Math in each section.

☐ Stop and do the Margin Exercises, including the Guided Solution Exercises, as directed.

STEP 2: PARTICIPATE: Making Connections through Active Exploration

☐ Explore the concepts using the Animations in MyLab Math.

☐ Work the Visualizing for Success or Translating for Success exercises in the text and/or in MyLab Math.

☐ Answer the Check Your Understanding exercises in the Section Exercises in the textbook and/or in MyLab Math.

STEP 3: PRACTICE: Reinforcing Understanding

☐ Complete your assigned homework from the textbook and/or in MyLab Math.

   ☐ When doing homework from the textbook, use the answer section to check your work.

   ☐ When doing homework in MyLab Math, use the Learning Aids, such as Help Me Solve This and View an Example, as needed, working toward being able to complete exercises without the aids.

☐ Work the exercises in the Mid-Chapter Review.

☐ Read the Study Guide and work the Review Exercises in the Chapter Summary and Review.

☐ Take the Chapter Test as a practice exam. To watch an instructor solve each problem, go to the Chapter Test Prep Videos in MyLab Math.

Use the *Studying for Success* tips in the text and the *MyLab Math Study Skills* modules (with videos) to help you develop effective time-management, note-taking, test-prep, and other skills.

Fraction, Decimal, Percent Equivalents

Fraction Notation	$\frac{1}{10}$	$\frac{1}{8}$	$\frac{1}{6}$	$\frac{1}{5}$	$\frac{1}{4}$	$\frac{3}{10}$	$\frac{1}{3}$	$\frac{3}{8}$	$\frac{2}{5}$	$\frac{1}{2}$	$\frac{3}{5}$	$\frac{5}{8}$	$\frac{2}{3}$	$\frac{7}{10}$	$\frac{3}{4}$	$\frac{4}{5}$	$\frac{5}{6}$	$\frac{7}{8}$	$\frac{9}{10}$	$\frac{1}{1}$
Decimal Notation	0.1	0.125	$0.16\overline{6}$	0.2	0.25	0.3	$0.33\overline{3}$	0.375	0.4	0.5	0.6	0.625	$0.66\overline{6}$	0.7	0.75	0.8	$0.83\overline{3}$	0.875	0.9	1
Percent Notation	10%	12.5% or $12\frac{1}{2}\%$	16.6% or $16\frac{2}{3}\%$	20%	25%	30%	33.3% or $33\frac{1}{3}\%$	37.5% or $37\frac{1}{2}\%$	40%	50%	60%	62.5% or $62\frac{1}{2}\%$	66.6% or $66\frac{2}{3}\%$	70%	75%	80%	83.3% or $83\frac{1}{3}\%$	87.5% or $87\frac{1}{2}\%$	90%	100%

Percent Notation

$$3\% = 3 \times \frac{1}{100} = \frac{3}{100} \qquad 3\% = 3 \times 0.01 = 0.03$$

Percent Increase and Decrease

Price increased from \$480 to \$504:

$$\frac{\text{Percent}}{\text{increase}} = \frac{\text{Change}}{\text{Original}} = \frac{504 - 480}{480} = \frac{24}{480} = 0.05 = 5\%$$

Price decreased from \$60 to \$45:

$$\frac{\text{Percent}}{\text{decrease}} = \frac{\text{Change}}{\text{Original}} = \frac{60 - 45}{60} = \frac{15}{60} = 0.25 = 25\%$$

Translating to Percent Equations

What is 20% of 75?
$$a = 20\% \cdot 75$$

15 is 20% of what?
$$15 = 20\% \cdot b$$

15 is what percent of 75?
$$15 = p \cdot 75$$

Interest

Simple: $I = P \cdot r \cdot t$

Compound: $A = P \cdot \left(1 + \dfrac{r}{n}\right)^{n \cdot t}$

American Units of Length

12 inches (in.) = 1 foot (ft)
36 inches = 1 yard (yd)
3 feet = 1 yard
5280 feet = 1 mile (mi)

Temperature

$$F = \frac{9}{5} \cdot C + 32 \qquad C = \frac{5}{9}(F - 32)$$

Metric Units of Length

1 kilometer (km) = 1000 meters (m)
1 hectometer (hm) = 100 meters
1 dekameter (dam) = 10 meters
1 meter
1 decimeter (dm) = $\dfrac{1}{10}$ meter
1 centimeter (cm) = $\dfrac{1}{100}$ meter
1 millimeter (mm) = $\dfrac{1}{1000}$ meter

Formulas

Perimeter of a Square: $P = 4 \cdot s$
Perimeter of a Rectangle: $P = 2 \cdot l + 2 \cdot w$
Area of a Square: $A = s^2$
Area of a Rectangle: $A = l \cdot w$
Area of a Parallelogram: $A = b \cdot h$
Area of a Triangle: $A = \dfrac{1}{2} \cdot b \cdot h$
Area of a Trapezoid: $A = \dfrac{1}{2} \cdot h \cdot (a + b)$
Circumference of a Circle: $C = \pi \cdot d$, or $C = 2 \cdot \pi \cdot r$
Area of a Circle: $A = \pi \cdot r^2$
Volume of a Rectangular Solid: $V = l \cdot w \cdot h$
Volume of a Circular Cylinder: $V = \pi \cdot r^2 \cdot h$

Pythagorean Theorem

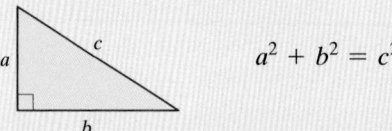

$$a^2 + b^2 = c^2$$

Exponents and Scientific Notation

$$x^0 = 1; \quad x^1 = x; \quad x^{-3} = \frac{1}{x^3}; \quad x^2 \cdot x^5 = x^7;$$

$$\frac{x^5}{x^2} = x^3; \quad (x^2)^5 = x^{10}; \quad (xy)^5 = x^5 y^5; \quad \left(\frac{x}{y}\right)^5 = \frac{x^5}{y^5}$$

$$3.26 \times 10^4 = 32{,}600$$

$$3.26 \times 10^{-4} = 0.000326$$

At a Glance: Prealgebra

Place Value

$$529{,}630{,}718{,}249.70651$$

Billions			Millions			Thousands			Ones							
5	2	9	6	3	0	7	1	8	2	4	9	7	0	6	5	1
Hundreds	Tens	Ones	Hundreds	Tens	Ones	Hundreds	Tens	Ones	Hundreds	Tens	Ones	Tenths	Hundredths	Thousandths	Ten-thousandths	Hundred-thousandths

Problem-Solving Strategy

1. Familiarize
2. Translate
3. Solve
4. Check
5. State

Exponential Notation

$$\text{Base} \to 3^{\overset{\text{Exponent}}{4}} = \underbrace{3 \cdot 3 \cdot 3 \cdot 3}_{4 \text{ factors of } 3}$$

Order of Operations

1. Do all calculations within grouping symbols before operations outside.
2. Evaluate all exponential expressions.
3. Do all multiplications and divisions in order from left to right.
4. Do all additions and subtractions in order from left to right.

Operations with Integers

$$-18 + 3 = -15 \qquad -9 \cdot 6 = -54$$
$$-6 + (-4) = -10 \qquad -5 \cdot (-3) = 15$$
$$9 - 12 = -3 \qquad 18 \div (-3) = -6$$
$$-7 - (-10) = 3 \qquad -10 \div (-2) = 5$$
$$\text{Absolute value: } |-4| = 4$$

Prime Factorization

$$220 = 2 \cdot 2 \cdot 5 \cdot 11 \qquad \text{All factors are prime.}$$

Least Common Multiple

$$12 = 2 \cdot 2 \cdot 3 \qquad 15 = 3 \cdot 5$$

The LCM of 12 and 15 $= 2 \cdot 2 \cdot 3 \cdot 5$, or 60.

Fraction Notation

$$\frac{8}{1} = 8 \qquad \frac{8}{8} = 1 \qquad \frac{0}{8} = 0 \qquad \frac{8}{0} \text{ Not defined}$$

$$\frac{5}{6} + \frac{3}{4} = \frac{5}{6} \cdot \frac{2}{2} + \frac{3}{4} \cdot \frac{3}{3} = \frac{10}{12} + \frac{9}{12} = \frac{10 + 9}{12} = \frac{19}{12}$$

$$\frac{5}{6} - \frac{3}{4} = \frac{5}{6} \cdot \frac{2}{2} - \frac{3}{4} \cdot \frac{3}{3} = \frac{10}{12} - \frac{9}{12} = \frac{10 - 9}{12} = \frac{1}{12}$$

$$\frac{5}{6} \cdot \frac{3}{4} = \frac{5 \cdot 3}{6 \cdot 4} = \frac{15}{24} = \frac{3 \cdot 5}{3 \cdot 8} = \frac{3}{3} \cdot \frac{5}{8} = 1 \cdot \frac{5}{8} = \frac{5}{8}$$

$$\frac{5}{6} \div \frac{3}{4} = \frac{5}{6} \cdot \frac{4}{3} = \frac{20}{18} = \frac{2 \cdot 10}{2 \cdot 9} = \frac{2}{2} \cdot \frac{10}{9} = 1 \cdot \frac{10}{9} = \frac{10}{9}$$

Mixed Numerals

$$\frac{11}{3} = 3\frac{2}{3} \qquad\qquad 4\frac{1}{2} = \frac{9}{2}$$

Graphing Linear Equations

x	$y = 2x - 3$	(x, y)
3	3	$(3, 3)$
1	-1	$(1, -1)$
0	-3	$(0, -3)$

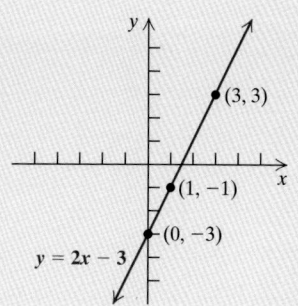

Student Organizer

Course Information

Course Number: _____ Name: _____

Location: _____ Days/Time: _____

Contact Information

Contact	Name	Email	Phone	Office Hours	Location
Instructor					
Tutor					
Math Lab					
Classmate					
Classmate					

Semester Organizer

Week	Homework	Quizzes and Tests	Other

EDITION

13

Intermediate Algebra

Marvin L. Bittinger

Indiana University Purdue University Indianapolis

Judith A. Beecher

Barbara L. Johnson

Ivy Tech Community College of Indiana

 Pearson

Director, Courseware Portfolio Management:	Michael Hirsch
Courseware Portfolio Manager:	Cathy Cantin
Courseware Portfolio Management Assistants:	Shannon Bushee; Shannon Slocum
Managing Producer:	Karen Wernholm
Content Producer:	Ron Hampton
Producer:	Erin Carreiro
Manager, Courseware QA:	Mary Durnwald
Manager, Content Development:	Eric Gregg
Field Marketing Managers:	Jennifer Crum; Lauren Schur
Product Marketing Manager:	Kyle DiGiannantonio
Product Marketing Assistant:	Brooke Imbornone
Senior Author Support/ Technology Specialist:	Joe Vetere
Manager, Rights and Permissions:	Gina Cheselka
Manufacturing Buyer:	Carol Melville, LSC Communications
Associate Director of Design:	Blair Brown
Program Design Lead:	Barbara T. Atkinson
Text Design:	Geri Davis/The Davis Group, Inc.
Editorial and Production Service:	Martha Morong/Quadrata, Inc.
Composition:	Cenveo® Publisher Services
Illustration:	Network Graphics; William Melvin
Cover Design:	Cenveo® Publisher Services
Cover Image:	Nick Veasey/Untitled X-Ray/Getty Images

Library of Congress Cataloging-in-Publication Data

Names: Bittinger, Marvin L., author. | Beecher, Judith A., author. | Johnson, Barbara L., author.
Title: Intermediate algebra / Marvin L. Bittinger (Indiana University Purdue University Indianapolis),
    Judith A. Beecher, Barbara L. Johnson (Ivy Tech Community College of Indiana).
Description: 13 edition. | New York, NY : Pearson, [2019] | Includes index.
Identifiers: LCCN 2017060142| ISBN 9780134707365 (pbk.) | ISBN 0134707362 (pbk.)
Subjects: LCSH: Algebra–Textbooks.
Classification: LCC QA154.3 .B578 2019 | DDC 512.9–dc23
LC record available at https://lccn.loc.gov/2017060142

12 2022

ISBN 13: 978-0-13-470736-5
ISBN 10: 0-13-470736-2

Contents

KD 12.19 2022 0831

Index of Activities

Index of Animations

Preface

Math doesn't change, but students' needs—and the way students learn—do.

With this in mind, *Intermediate Algebra,* 13th Edition, continues the Bittinger tradition of objective-based, guided learning, while integrating many updates with the proven pedagogy. These updates are motivated by feedback that we received from students and instructors, as well as our own experience in the classroom. In this edition, our focus is on guided learning and retention: helping each student (and instructor) get the most out of all the available program resources—wherever and whenever they engage with the math.

We believe that student success in math hinges on four key areas: **Foundation, Engagement, Application,** and **Retention**. In the 13th edition, we have added key new program features (highlighted below, for quick reference) in each area to make it easier for each student to personalize his or her learning experience. In addition, you will recognize many proven features and presentations from the previous edition of the program.

FOUNDATION
Studying the Concepts

Students can learn the math concepts by reading the textbook or the eText, participating in class, watching the videos, working in the *MyMathGuide* workbook—or using whatever combination of these course resources works best for them.

In order to understand new math concepts, students must recall and use skills and concepts previously studied. To support student learning, we have integrated two important new features throughout the 13th Edition program:

☐ *New!* **Just-in-Time Review** at the beginning of the text and the eText is a set of quick reviews of the key topics from previous courses that are prerequisites for the new material in this course. A note on each Chapter Opener alerts students to the topics they should review for that chapter. In MyLab Math, students will find a concise presentation of each topic in the **Just-in-Time Review Videos.**

☐ *New!* **Skill Review,** in nearly every section of the text and the eText, reviews a previously presented skill at the objective level where it is key to learning the new material. This feature offers students two practice exercises with answers. In MyLab Math, new **Skill Review Videos,** created by the Bittinger author team, offer a concise, step-by-step solution for each Skill Review exercise.

Margin Exercises with Guided Solutions, with fill-in blanks at key steps in the problem-solving process, appear in nearly every text section and can be assigned in MyLab Math.

Algebraic–Graphical Connections in the text draw explicit connections between the algebra and the corresponding graphical visualization.

Intermediate Algebra Video Program, our comprehensive program of objective-based, interactive videos, can be used hand-in-hand with our *MyMathGuide* workbook. **Interactive Your Turn exercises** in the videos prompt students to solve problems and receive instant feedback. These videos can be accessed at the section, objective, and example levels.

MyMathGuide offers students a guided, hands-on learning experience. This objective-based workbook (available in print and in MyLab Math) includes vocabulary, skill, and concept review—as well as problem-solving practice with space for students to fill in the answers and stepped-out solutions to problems, to show (and keep) their work, and to write notes. Students can use *MyMathGuide,* while watching the videos, listening to the instructor's lecture, or reading the text or the eText, in order to reinforce and self-assess their learning.

Studying for Success sections are checklists of study skills designed to ensure that students develop the skills they need to succeed in math, school, and life. They are available at the beginning of selected sections.

ENGAGEMENT
Making Connections through Active Exploration

Since understanding the big picture is key to student success, we offer many active learning opportunities for the practice, review, and reinforcement of important concepts and skills.

- ☐ *New!* **Chapter Opener Applications** with graphics use current data and applications to present the math in context. Each application is related to exercises in the text to help students model, visualize, learn, and retain the math.

- ☐ *New!* **Student Activities,** included with each chapter, have been developed as multistep, data-based activities for students to apply the math in the context of an authentic application. Student Activities are available in *MyMathGuide* and in MyLab Math.

- ☐ *New!* **Interactive Animations** can be manipulated by students in MyLab Math through guided and open-ended exploration to further solidify their understanding of important concepts.

Translating for Success offers extra practice with the important first step of the process for solving applied problems. **Visualizing for Success** asks students to match an equation or an inequality with its graph by focusing on characteristics of the equation or the inequality and the corresponding attributes of the graph. Both of these activities are available in the text and in MyLab Math.

Calculator Corner is an optional feature in each chapter that helps students use a calculator to perform calculations and to visualize concepts.

Learning Catalytics uses students' mobile devices for an engagement, assessment, and classroom intelligence system that gives instructors real-time feedback on student learning.

APPLICATION
Reinforcing Understanding

As students explore the math, they have frequent opportunities to apply new concepts, practice, self-assess, and reinforce their understanding.

Margin Exercises, labeled "Do Exercise . . . ," give students frequent opportunities to apply concepts just discussed by solving problems that parallel text examples.

Exercise Sets in each section offer abundant opportunity for practice and review in the text and in MyLab Math. The Section Exercises are grouped by objective for ease of use, and each set includes the following special exercise types:

- ☐ *New!* **Check Your Understanding** with **Reading Check** and **Concept Check** exercises, at the beginning of each exercise set, gives students the opportunity to assess their grasp of the skills and concepts before moving on to the objective-based section exercises. In MyLab Math, many of these exercises use drag & drop functionality.
- ☐ **Skill Maintenance Exercises** offer a thorough review of the math in the preceding sections of the text.
- ☐ **Synthesis Exercises** help students develop critical-thinking skills by requiring them to use what they know in combination with content from the current and previous sections.

RETENTION
Carrying Success Forward

Because continual practice and review is so important to retention, we have integrated both throughout the program in the text and in MyLab Math.

- ☐ *New!* **Skill Builder** adaptive practice, available in MyLab Math, offers each student a personalized learning experience. When a student struggles with the assigned homework, Skill Builder exercises offer just-in-time additional adaptive practice. The adaptive engine tracks student performance and delivers to each individual questions that are appropriate for his or her level of understanding. When the system has determined that the student has a high probability of successfully completing the assigned exercise, it suggests that the student return to the assigned homework.

Mid-Chapter Review offers an opportunity for active review midway through each chapter. This review offers four types of practice problems:

Concept Reinforcement, Guided Solutions, Mixed Review, and Understanding Through Discussion and Writing

Summary and Review is a comprehensive learning and review section at the end of each chapter. Each of the five sections—**Vocabulary Reinforcement** (fill-in-the-blank), **Concept Reinforcement** (true/false), **Study Guide** (examples with stepped-out solutions paired with similar practice problems), **Review Exercises**, and **Understanding Through Discussion and Writing**—includes references to the section in which the material was covered to facilitate review.

Chapter Test offers students the opportunity for comprehensive review and reinforcement prior to taking their instructor's exam. **Chapter Test Prep Videos** in MyLab Math show step-by-step solutions to the questions on the chapter test.

Cumulative Review follows each chapter beginning with Chapter 2. These revisit skills and concepts from all preceding chapters to help students retain previously presented material.

Resources for Success

MyLab Math Online Course for Bittinger, Beecher, and Johnson, *Intermediate Algebra*, 13th Edition (access code required)

MyLab™ Math is available to accompany Pearson's market-leading text offerings. To give students a consistent tone, voice, and teaching method, the pedagogical approach of the text is tightly integrated throughout the accompanying MyLab Math course, making learning the material as seamless as possible.

UPDATED! Learning Path

Structured, yet flexible, the updated learning path highlights author-created, faculty-vetted content—giving students what they need exactly when they need it. The learning path directs students to resources such as two new types of video: **Just-in-Time Review** (concise presentations of key topics from previous courses) and **Skill Review** (author-created exercises with step-by-step solutions that reinforce previously presented skills), both available in the Multimedia Library and assignable in MyLab Math.

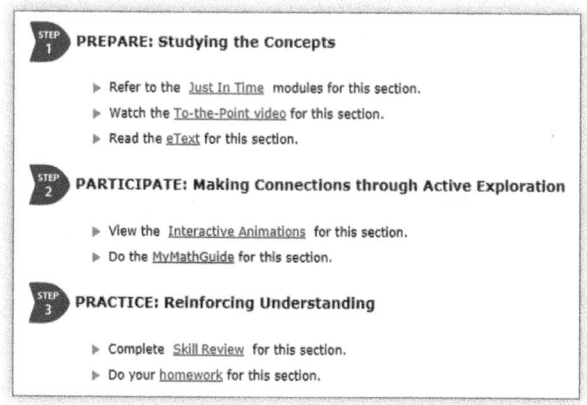

STEP 1 **PREPARE: Studying the Concepts**
- ▶ Refer to the Just In Time modules for this section.
- ▶ Watch the To-the-Point video for this section.
- ▶ Read the eText for this section.

STEP 2 **PARTICIPATE: Making Connections through Active Exploration**
- ▶ View the Interactive Animations for this section.
- ▶ Do the MyMathGuide for this section.

STEP 3 **PRACTICE: Reinforcing Understanding**
- ▶ Complete Skill Review for this section.
- ▶ Do your homework for this section.

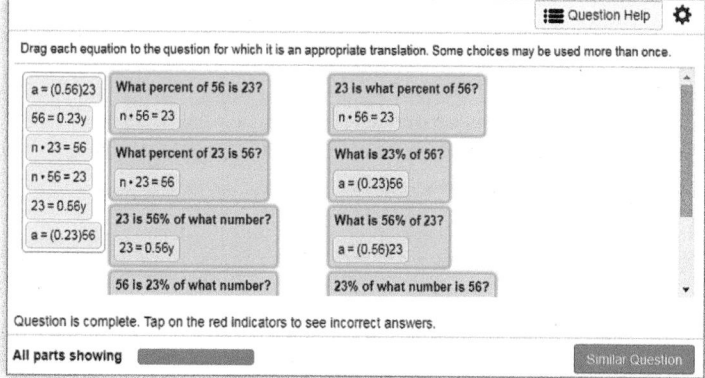

NEW!
Drag-and-Drop Exercises

Drag-and-drop exercises are now available in MyLab Math. This new assignment type allows students to drag answers and values within a problem, providing a new and engaging way to test students' concept knowledge.

NEW and UPDATED! Animations

New animations encourage students to learn key concepts through guided and open-ended exploration. Animations are available through the learning path and multimedia library, and they can be assigned within MyLab Math.

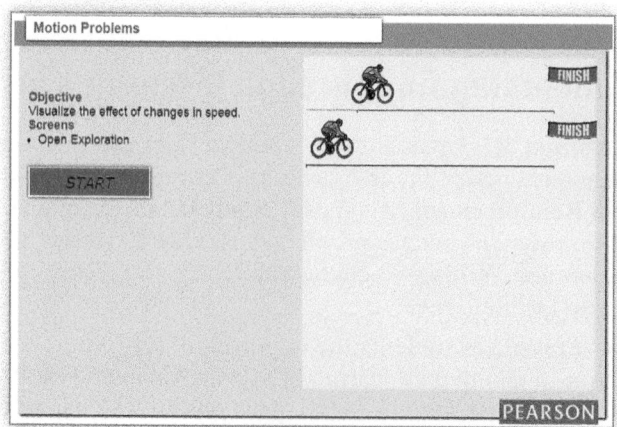

pearson.com/mylab/math

Resources for Success

Instructor Resources

Additional resources can be downloaded from **www.pearsohhighered.com** or hardcopy resources can be ordered from your sales representative.

Annotated Instructor's Edition

ISBN: 0134708334

- Answers to all text exercises.
- Helpful teaching tips, including suggestions for incorporating Student Activities in the course

Instructor's Resource Manual with Tests and Minilectures

(download only)
ISBN: 0134719042

- Resources designed to help both new and experienced instructors with course preparation and class management.
- Chapter teaching tips and support for media supplements.
- Multiple versions of multiple-choice and free-response chapter tests, as well as final exams.

Instructor's Solutions Manual

(download only)
By Judy Penna
ISBN: 0134708350
The *Instructor's Solutions Manual* includes brief solutions for the even-numbered exercises in the exercise sets and fully worked-out annotated solutions for all the exercises in the Mid-Chapter Reviews, the Summary and Reviews, the Chapter Tests, and the Cumulative Reviews.

PowerPoint® Lecture Slides

(download only)

- Editable slides present key concepts and definitions from the text.
- Available to both instructors and students.
- Fully accessible.

TestGen®

TestGen enables instructors to build, edit, print, and administer tests using a computerized test bank of questions developed to cover all the objectives of the text. (www.pearsoned.com/testgen)

Student Resources

Intermediate Algebra Lecture Videos

- Concise, interactive, and objective-based videos.
- View a whole section, choose an objective, or go straight to an example.

Chapter Test Prep Videos

- Step-by-step solutions for every problem in the chapter tests.

Just-in-Time Review Videos

- One video per review topic in the Just-in-Time Review at the beginning of the text.
- View examples and worked-out solutions that parallel the concepts reviewed in each review topic.

Skill Review Videos

Students can review previously presented skills at the objective level with two practice exercises before moving forward in the content. Videos include a step-by-step solution for each exercise.

MyMathGuide: Notes, Practice, and Video Path

ISBN: 0134708385

- Guided, hands-on learning in a workbook format with space for students to show their work and record their notes and questions.
- Highlights key concepts, skills, and definitions; offers quick reviews of key vocabulary terms with practice problems, examples with guided solutions, similar Your Turn exercises, and practice exercises with readiness checks.
- Includes student activities utilizing real data.
- Available in MyLab Math and as a printed manual.

Student's Solutions Manual

ISBN: 0134719069
By Judy Penna

- Includes completely worked-out annotated solutions for odd-numbered exercises in the text, as well as all the exercises in the Mid-Chapter Reviews, the Summary and Reviews, the Chapter Tests, and the Cumulative Reviews.
- Available in MyLab Math and as a printed manual.

pearson.com/mylab/math

Acknowledgments

Our deepest appreciation to all the instructors and students who helped to shape this revision of our program by reviewing our texts and courses, providing feedback, and sharing their experiences with us at conferences and on campus. In particular, we would like to thank the following for reviewing the titles in our worktext program for this revision:

Amanda L. Blaker, *Gallatin College*
Jessica Bosworth, *Nassau Community College*
Judy G. Burn, *Trident Technical College*
Abushieba A. Ibrahim, *Nova Southeastern University*
Laura P. Kyser, *Savannah Technical College*
David Mandelbaum, *Nova Southeastern University*

An outstanding team of professionals was involved in the production of this text. We want to thank Judy Penna for creating the new Skill Review videos and for writing the *Student's Solutions Manual* and the *Instructor's Solutions Manual.* We also thank Laurie Hurley for preparing *MyMathGuide*, Robin Rufatto for creating the new Just-in-Time videos, and Tom Atwater for supporting and overseeing the new videos. Accuracy checkers Judy Penna, Laurie Hurley, and Susan Meshulam contributed immeasurably to the quality of the text.

Martha Morong, of Quadrata, Inc., provided editorial and production services of the highest quality, and Geri Davis, of The Davis Group, performed superb work as designer, art editor, and photo researcher. Their countless hours of work and consistent dedication have led to products of which we are immensely proud.

In addition, a number of people at Pearson, including the Developmental Math Team, have contributed in special ways to the development and production of our program. Special thanks are due to Cathy Cantin, Courseware Portfolio Manager, for her visionary leadership and development support. In addition, Ron Hampton, Content Producer, contributed invaluable coordination for all aspects of the project. We also thank Erin Carreiro, Producer, and Kyle DiGiannantonio, Product Marketing Manager, for their exceptional support.

Our goal in writing this textbook was to make mathematics accessible to every student. We want you to be successful in this course and in the mathematics courses you take in the future. Realizing that your time is both valuable and limited, and that you learn in a uniquely individual way, we employ a variety of pedagogical and visual approaches to help you learn in the best and most efficient way possible. We wish you a positive and successful learning experience.

Marv Bittinger
Judy Beecher
Barbara Johnson

Index of Applications

The common octopus, *Octopus vulgaris*, is most often found in temperate and tropical waters at a depth of 100 to 150 meters. This species can reach a length of 1 meter including its legs. The octopus has three hearts and no bones. Octopuses can learn and are probably the most intelligent of all invertebrates. One-third of its neurons are located in its brain, and the remaining two-thirds of its neurons are in its arms, which are lined with suckers. The table at left shows a comparison of the number of neurons in a human, an octopus, a mouse, and a pond snail.

Comparing Number of Neurons

	Number of neurons
Pond snail	10,000
Mouse	80,000,000
Octopus	500,000,000
Human	86,000,000,000

DATA: *National Geographic*, November 2016, p.70; "The Power of Eight," by Olivia Judson

Data: animaldiversity.org; arkive.org; National Geographic, November 2016, p. 70, "The Power of Eight," Olivia Judson, Roger Hanlon, Marine Biological Laboratory; Guy Levy and Benny Hochner, Hebrew University of Jerusalem; Cliff Ragsdale, University of Chicago

We will calculate the number of neurons in the arms of an *Octopus vulgaris* in Example 5 of Just-in-Time 20.

Just-in-Time Review

MyLab Math
VIDEO

The following figure shows the relationships among various sets of real numbers.

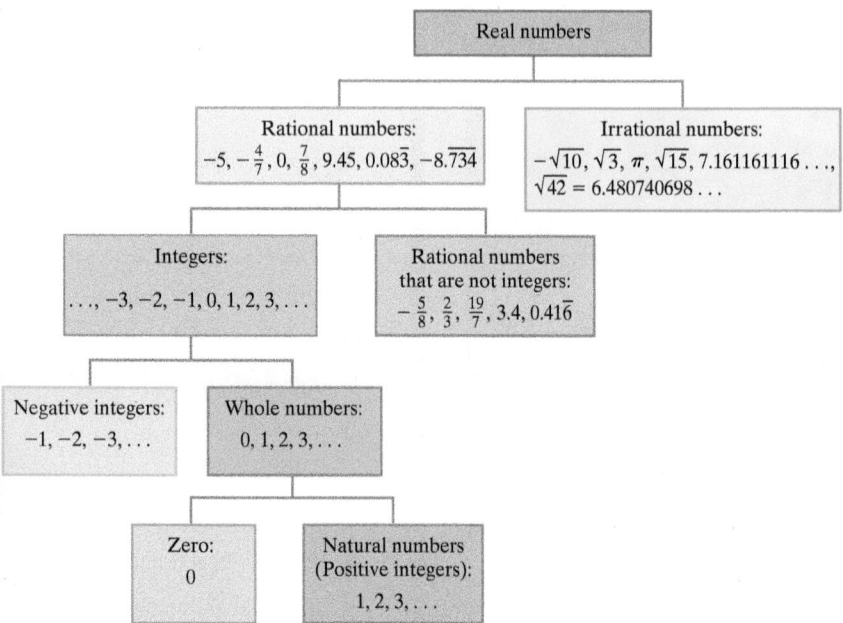

The set containing the numbers $-5, 0,$ and 3 can be named $\{-5, 0, 3\}$. It is described using the **roster method**, which lists all members of a set. We use the roster method to describe three frequently used subsets of real numbers. Note that three dots are used to indicate that the pattern continues without end.

Natural numbers: $\{1, 2, 3, \ldots\}$

Whole numbers: $\{0, 1, 2, 3, \ldots\}$

Integers: $\{\ldots, -3, -2, -1, 0, 1, 2, 3, \ldots\}$

Other subsets of real numbers are described using **set-builder notation**. With this notation, instead of listing all members of a set, we specify conditions under which a number is in a set. For example, the set of all odd natural numbers less than 9 can be described as follows:

$\{x \mid x$ is an odd number less than $9\}$, and read

The set of all x such that x is an odd number less than 9.

Using roster notation, we can write this set as $\{1, 3, 5, 7\}$.

EXERCISES

1. Use roster notation to name each set.

   a) The set of all letters in the word "solve"

   b) The set of all even natural numbers

2. Use set-builder notation to name each set.

   a) $\{0, 1, 2, 3, 4, 5\}$

   b) $\{4, 5, 6, 7, 8, 9, 10\}$

(continued)

The set of **real numbers** is

$$\{x \mid x \text{ is a rational number } or \ x \text{ is an irrational number}\}.$$

Every point on the number line represents some real number, and every real number is represented by some point on the number line.

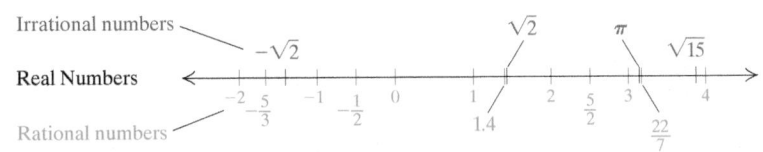

A **rational number** can be expressed as an integer divided by a nonzero integer. The set of rational numbers is

$$\left\{ \frac{p}{q} \ \middle| \ p \text{ is an integer, } q \text{ is an integer, and } q \neq 0 \right\}.$$

Rational numbers are numbers whose decimal representation either terminates or has a repeating block of digits. The following are examples of rational numbers:

$$\frac{5}{8}, \quad \frac{12}{-7}, \quad \frac{-17}{15}, \quad -\frac{9}{7}, \quad \frac{39}{1}, \quad \frac{0}{6}.$$

Note that $\frac{39}{1} = 39$. Thus the set of rational numbers contains the integers. Using long division, we can write a fraction in decimal notation:

$$\frac{5}{8} = \underbrace{0.625}_{\text{Terminating}} \quad \text{and} \quad \frac{6}{11} = \underbrace{0.545454\ldots}_{\text{Repeating}} = 0.\overline{54}.$$

The bar in $0.\overline{54}$ indicates the repeating block of digits in decimal notation.

Irrational numbers are numbers whose decimal representation neither terminates nor has a repeating block of digits. They cannot be represented as the quotient of two integers. Numbers like π, $\sqrt{2}$, $-\sqrt{10}$, $\sqrt{13}$, and $-1.898898889\ldots$ are examples of irrational numbers. The decimal notation for an irrational number *neither* terminates *nor* repeats.

Do Exercises 1–4.
(Exercises 1 and 2 are on the preceding page.) ▶

(Exercises 1 and 2 are on the preceding page.)

EXERCISES

3. Given the numbers

$$20, \ -10, \ -5.34, \ 18.999, \ 0.58\overline{3},$$
$$\frac{11}{45}, \ \sqrt{7}, \ -\sqrt{2}, \ \sqrt{16}, \ 0, \ -\frac{2}{3},$$
$$9.34334333433334\ldots:$$

a) Name the natural numbers.
b) Name the whole numbers.
c) Name the integers.
d) Name the irrational numbers.
e) Name the rational numbers.
f) Name the real numbers.

4. Given the numbers

$$-6, \ 0, \ 1, \ -\frac{1}{2}, \ -4, \ \frac{7}{9},$$
$$12, \ -\frac{6}{5}, \ 3.45, \ 5\frac{1}{2}, \ \sqrt{3},$$
$$\sqrt{25}, \ -\frac{12}{3}, \ 0.131331333133331\ldots$$

a) Name the natural numbers.
b) Name the whole numbers.
c) Name the integers.
d) Name the irrational numbers.
e) Name the rational numbers.
f) Name the real numbers.

Just-in-Time Review

2 ▸ ORDER FOR THE REAL NUMBERS

MyLab Math
VIDEO

Real numbers are named in order on the number line. For any two numbers on the line, the one to the left is less than the one to the right.

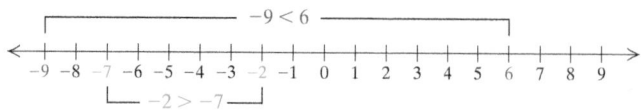

$$-9 < 6$$
$$-2 > -7$$

We use the symbol **<** to mean "**is less than.**" The sentence $-9 < 6$ means "-9 is less than 6." The symbol **>** means "**is greater than.**" The sentence $-2 > -7$ means "-2 is greater than -7." Sentences containing $<$ or $>$ are called **inequalities**.

EXAMPLES Use either $<$ or $>$ for ☐ to write a true sentence.

1. $4 \square 9$ Since 4 is to the left of 9, 4 is less than 9, so $4 < 9$.

2. $-8 \square 3$ Since -8 is to the left of 3, we have $-8 < 3$.

3. $7 \square -12$ Since 7 is to the right of -12, then $7 > -12$.

4. $-21 \square -5$ Since -21 is to the left of -5, we have $-21 < -5$.

5. $4.79 \square 4.97$ Since 4.79 is to the left of 4.97, we have $4.79 < 4.97$.

6. $-2.7 \square -\dfrac{3}{2}$ Since $-\dfrac{3}{2} = -1.5$ and -2.7 is to the left of -1.5, we have $-2.7 < -\dfrac{3}{2}$.

7. $\dfrac{5}{8} \square \dfrac{7}{11}$ We convert to decimal notation $\left(\dfrac{5}{8} = 0.625 \text{ and } \dfrac{7}{11} = 0.6363\ldots\right)$ and compare. Thus, $\dfrac{5}{8} < \dfrac{7}{11}$. ■

Note that $-8 < 5$ and $5 > -8$ are both true. Every true inequality yields another true inequality if we interchange the numbers or variables and reverse the direction of the inequality sign.

$a < b$ also has the meaning $b > a$.

EXERCISES

Insert $<$ or $>$ for ☐ to write a true sentence.

1. $-5 \square -4$

2. $-\dfrac{1}{4} \square -\dfrac{1}{2}$

3. $87 \square 67$

4. $-9.8 \square -4\dfrac{2}{3}$

5. $6.78 \square -6.77$

6. $-\dfrac{4}{5} \square -0.86$

7. $\dfrac{14}{29} \square \dfrac{17}{32}$

8. $-\dfrac{12}{13} \square -\dfrac{14}{15}$

9. $1.8 \square 1.08$

10. $0 \square -4$

MyLab Math
ANIMATION

(continued)

EXAMPLES Write a different inequality with the same meaning.

8. $a < -5$ The inequality $-5 > a$ has the same meaning.

9. $-3 > -8$ The inequality $-8 < -3$ has the same meaning.

Expressions like $a \le b$ and $a \ge b$ are also **inequalities**. We read $a \le b$ as "**a is less than or equal to b.**" We read $a \ge b$ as "**a is greater than or equal to b.**" If a is nonnegative, then $a \ge 0$.

EXAMPLES Determine whether each of the following is true or false.

10. $-8 \le 5.7$ True since $-8 < 5.7$ is true.

11. $-8 \le -8$ True since $-8 = -8$ is true.

12. $-7 \ge 4\frac{1}{3}$ False since neither $-7 > 4\frac{1}{3}$ nor $-7 = 4\frac{1}{3}$ is true.

13. $-\frac{2}{3} \ge -\frac{5}{4}$ True since $-\frac{2}{3} = -0.666\ldots$ and $-\frac{5}{4} = -1.25$ and $-0.666\ldots > -1.25$.

Do Exercises 1–16.
(Exercises 1–10 are on the preceding page.) ▶

EXERCISES

Write a different inequality with the same meaning.

11. $x > 6$

12. $-4 < 7$

Determine whether each of the following is true or false.

13. $6 \ge -9.4$

14. $-18 \le -18$

15. $-7.6 \le -10\frac{4}{5}$

16. $-\frac{24}{27} \ge -\frac{25}{28}$

MyLab Math

VIDEO

A replacement that makes an inequality true is called a **solution**. The set of all solutions is called the **solution set**. A **graph** of an inequality is a drawing that represents its solution set.

EXAMPLE 1 Graph the inequality $x > -3$ on the number line.

The solutions consist of all real numbers greater than -3, so we shade all numbers greater than -3. Since -3 is not a solution, we use a parenthesis at -3 to indicate this. The graph represents the solution set $\{x \mid x > -3\}$.

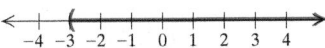

EXAMPLE 2 Graph the inequality $x \le 2$ on the number line.

We make a drawing that represents the solution set $\{x \mid x \le 2\}$. The graph consists of 2 as well as the numbers less than 2. We shade all numbers to the left of 2 and use a bracket at 2 to indicate that it is also a solution.

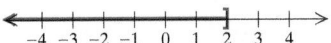

Do Exercises 1–4. ▶

EXERCISES

Graph each inequality.

1. $x > -1$

2. $x \le 5$

3. $x > 0$

4. $x \le -\dfrac{5}{2}$

4 ABSOLUTE VALUE

We call the distance of a number from 0 on the number line the **absolute value** of the number. Since distance is always a nonnegative number, the absolute value of a number is always greater than or equal to 0.

The distance from -6 to 0 is 6.
The absolute value of -6 is 6.

The distance from 6 to 0 is 6.
The absolute value of 6 is 6.

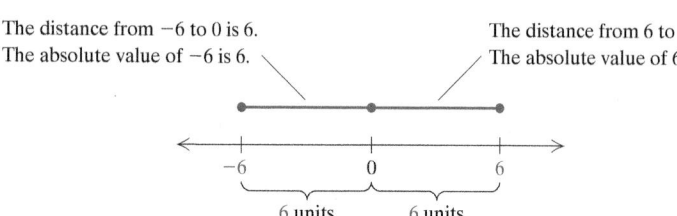

6 units   6 units

We use the symbol $|x|$ to represent the absolute value of a number x.

EXAMPLES Find the absolute value.

1. $|-7|$ The distance of -7 from 0 is 7, so $|-7|$ is 7.
2. $|12|$ The distance of 12 from 0 is 12, so $|12|$ is 12.
3. $|0|$ The distance of 0 from 0 is 0, so $|0|$ is 0.
4. $\left|\dfrac{4}{5}\right| = \dfrac{4}{5}$
5. $|-3.86| = 3.86$

Do Exercises 1–6. ▶

EXERCISES

Find the absolute value.

1. $|-4|$

2. $|0|$

3. $|252.7|$

4. $|31|$

5. $\left|-\dfrac{1}{2}\right|$

6. $|-0.03|$

RULES FOR ADDITION OF REAL NUMBERS

1. *Positive numbers*: Add the numbers. The result is positive.
2. *Negative numbers*: Add absolute values. Make the answer negative.
3. *A positive number and a negative number*:
   - If the numbers have the same absolute value, the answer is 0.
   - If the numbers have different absolute values, subtract the smaller absolute value from the larger. Then:
     a) If the positive number has the greater absolute value, make the answer positive.
     b) If the negative number has the greater absolute value, make the answer negative.
4. *One number is zero*: The sum is the other number.

IDENTITY PROPERTY OF ZERO

For any real number a, $a + 0 = a$.
(The number 0 is the **additive identity**.)

EXAMPLES

1. $-13 + (-8) = -21$ Two negatives. Add the absolute values: $|-13| + |-8| = 13 + 8 = 21$. Make the answer *negative*: -21.

2. $-2.1 + 8.5 = 6.4$ One negative, one positive. Subtract the smaller absolute value from the larger: $8.5 - 2.1 = 6.4$. The *positive* number, 8.5, has the larger absolute value, so the answer is *positive*, 6.4.

3. $-48 + 31 = -17$ One negative, one positive. Subtract the smaller absolute value from the larger: $48 - 31 = 17$. The *negative* number, -48, has the larger absolute value, so the answer is *negative*, -17.

4. $2.6 + (-2.6) = 0$ One positive, one negative. The numbers have the same absolute value. The sum is 0.

5. $-\dfrac{5}{9} + 0 = -\dfrac{5}{9}$ One number is zero. The sum is $-\frac{5}{9}$.

6. $-\dfrac{2}{3} + \dfrac{5}{8} = -\dfrac{16}{24} + \dfrac{15}{24} = -\dfrac{1}{24}$

Do Exercises 1–14. ▶

EXERCISES

Add.

1. $8 + (-3)$

2. $-8 + 3$

3. $-8 + (-3)$

4. $8 + 3$

5. $7 + (-2)$

6. $-8 + (-8)$

7. $-13 + 13$

8. $-26 + 0$

9. $-6 + (-15)$

10. $9 + (-2)$

11. $-8.4 + 9.6$

12. $-2.62 + (-6.24)$

13. $-\dfrac{2}{5} + \dfrac{3}{4}$

14. $-\dfrac{5}{6} + \left(-\dfrac{7}{8}\right)$

6 ▸ OPPOSITES, OR ADDITIVE INVERSES

MyLab Math

VIDEO

Two numbers whose sum is 0 are called **opposites**, or **additive inverses**.

EXAMPLES Find the opposite, or additive inverse, of each number.

1. 8.6   The opposite of 8.6 is -8.6 because $8.6 + (-8.6) = 0$.

2. 0   The opposite of 0 is 0 because $0 + 0 = 0$.

3. $-\frac{7}{9}$ The opposite of $-\frac{7}{9}$ is $\frac{7}{9}$ because $-\frac{7}{9} + \frac{7}{9} = 0$.

To name the opposite, or additive inverse, we use the symbol $-$, and read the symbolism $-a$ as "the opposite of a" or "the additive inverse of a."

OPPOSITES, OR ADDITIVE INVERSES

For any real number a, the **opposite**, or **additive inverse**, of a, which is $-a$, is such that

$$a + (-a) = (-a) + a = 0.$$

EXAMPLES Evaluate $-x$ and $-(-x)$.

4. If $x = 23$, then $-x = -(23) = -23$ and

$$-(-x) = -(-23) = 23.$$

5. If $x = -5$, then $-x = -(-5) = 5$ and

$$-(-x) = -(-(-5)) = -(5) = -5.$$

Signs of Numbers

We sometimes say that a negative number has a "negative sign," and a positive number has a "positive sign." When we replace a number with its opposite, or additive inverse, we can say that we have "changed its sign."

EXAMPLES Change the sign. (Find the opposite, or additive inverse.)

6. -3 $-(-3) = 3$ **7.** $-\frac{3}{8}$ $-\left(-\frac{3}{8}\right) = \frac{3}{8}$

8. 0 $-0 = 0$ **9.** 14 $-(14) = -14$

We can now give a more formal definition of absolute value. For any real number a, the **absolute value** of a, denoted $|a|$, is given by

$$|a| = \begin{cases} a, & \text{if } a \geq 0, \\ -a, & \text{if } a < 0. \end{cases} \quad \text{For example, } |8| = 8 \text{ and } |0| = 0.$$
$$\text{For example, } |-5| = -(-5) = 5.$$

(The absolute value of a is a if a is nonnegative. The absolute value of a is the opposite of a if a is negative.)

Do Exercises 1–6. ▶

EXERCISES

Find the opposite (additive inverse).

1. 10

2. $-\frac{2}{3}$

3. -0.007

4. 0

5. If $x = 9$, evaluate $-x$ and $-(-x)$.

6. If $x = -23$, evaluate $-x$ and $-(-x)$.

·········· **Caution!** ··········

A symbol such as -8 is usually read "negative 8." It could be read "the opposite of 8," because the opposite of 8 is -8. It could also be read "the additive inverse of 8," because the additive inverse of 8 is -8. When a variable is involved, as in a symbol like $-x$, it can be read "the opposite of x" or "the additive inverse of x" but *not* "negative x," because we do not know whether the symbol represents a positive number, a negative number, or 0. It is never correct to read -8 as "minus 8."

The difference $a - b$ is the unique number c for which $a = b + c$. That is, $a - b = c$ if c is the number such that $a = b + c$. For example, $3 - 5 = -2$ because $3 = 5 + (-2)$. That is, -2 is the number that when added to 5 gives 3. Although this illustrates the formal definition of subtraction, we generally use the following when we subtract.

SUBTRACTION OF REAL NUMBERS

For any real numbers a and b,

$$a - b = a + (-b).$$

(We can subtract by adding the opposite (additive inverse) of the number being subtracted.)

Compare the following:

$$9 - 4\ = 9 + (-4) = 5;$$
$$9 - (-4) = 9 + 4 = 13;$$
$$-9 - 4\ = -9 + (-4) = -13;\quad \text{and}$$
$$-9 - (-4) = -9 + 4 = -5.$$

EXAMPLES Subtract.

1. $3 - 5 = 3 + (-5) = -2$

2. $7 - (-3) = 7 + 3 = 10$

3. $0 - 8 = 0 + (-8) = -8$

4. $-19.4 - 5.6 = -19.4 + (-5.6) = -25$

5. $-\dfrac{4}{3} - \left(-\dfrac{2}{5}\right) = -\dfrac{4}{3} + \dfrac{2}{5} = -\dfrac{20}{15} + \dfrac{6}{15} = -\dfrac{14}{15}$

Do Exercises 1–14. ▶

EXERCISES

Subtract.

1. $5 - 11$

2. $5 - (-11)$

3. $-5 - 11$

4. $-5 - (-11)$

5. $17 - (-17)$

6. $-6 - 14$

7. $-9 - (-9)$

8. $8 - 13$

9. $31 - (-16)$

10. $15.8 - 27.4$

11. $-18.01 - 11.24$

12. $\dfrac{5}{6} - \left(-\dfrac{1}{12}\right)$

13. $-\dfrac{1}{3} - \left(-\dfrac{1}{12}\right)$

14. $\dfrac{1}{3} - \dfrac{4}{5}$

8 ▶ MULTIPLY REAL NUMBERS

Look for a pattern:

$$3 \cdot 5 = 15 \qquad -2 \cdot 5 = -10$$
$$2 \cdot 5 = 10 \qquad -3 \cdot 5 = -15$$
$$1 \cdot 5 = 5 \qquad -4 \cdot 5 = -20$$
$$0 \cdot 5 = 0 \qquad -5 \cdot 5 = -25$$
$$-1 \cdot 5 = -5 \qquad -6 \cdot 5 = -30$$

Look for a pattern:

$$4(-5) = -20 \qquad -1(-5) = 5$$
$$3(-5) = -15 \qquad -2(-5) = 10$$
$$2(-5) = -10 \qquad -3(-5) = 15$$
$$1(-5) = -5 \qquad -4(-5) = 20$$
$$0(-5) = 0 \qquad -5(-5) = 25$$

RULES FOR MULTIPLICATION OF REAL NUMBERS
To multiply two real numbers:
1. If the signs are the same, multiply the absolute values and make the answer *positive*.
2. If the signs are different, multiply the absolute values and make the answer *negative*.
3. If one number is 0, the product is 0.

EXAMPLES   Multiply.

1. $-3 \cdot 5 = -15$ The signs are different. Multiply the absolute values. The answer is negative.

2. $(-8.8)(-3.5) = 30.8$ The signs are the same. Multiply the absolute values. The answer is positive.

3. $3 \cdot \left(-\frac{1}{2}\right) = \frac{3}{1} \cdot \left(-\frac{1}{2}\right) = -\frac{3}{2}$

4. $-3 \cdot (-5) = 15$

5. $(-1.2)(4.5) = -5.4$

6. $-\frac{7}{8} \cdot 0 = 0$

7. $\left(-\frac{3}{4}\right) \cdot \left(-\frac{5}{2}\right) = \frac{15}{8}$

Do Exercises 1–10. ▶

EXERCISES
Multiply.

1. $3(-7)$

2. $(-4.2)(-6.3)$

3. $-5 \cdot 9$

4. $-\frac{13}{7} \cdot \left(-\frac{5}{2}\right)$

5. $0 \cdot (-11)$

6. $-\frac{3}{5} \cdot \frac{4}{7}$

7. $-4(-13)$

8. $-3\left(\frac{2}{3}\right)$

9. $-\frac{9}{11} \cdot \left(-\frac{11}{9}\right)$

10. $-3(-4)(5)$

The quotient $a \div b$, or a/b, where $b \neq 0$, is that unique real number c for which $a = b \cdot c$. Using this definition and the rules for multiplying, we can see how to handle signs when dividing.

EXAMPLES Divide.

1. $\dfrac{10}{-2} = -5$, because $-2 \cdot (-5) = 10$

2. $\dfrac{-32}{4} = -8$, because $-8 \cdot (4) = -32$

3. $-25 \div (-5) = 5$, because $5 \cdot (-5) = -25$

RULES FOR DIVISION OF REAL NUMBERS

To divide two real numbers:

1. If the signs are the same, divide the absolute values and make the answer *positive*.

2. If the signs are different, divide the absolute values and make the answer *negative*.

Excluding Division by Zero

We cannot divide a nonzero number n by zero. By the definition of division, $n/0$ would be some number that when multiplied by 0 gives n. But when any number is multiplied by 0, the result is 0. Thus the only possibility for n would be 0.

Consider $0/0$. Using the definition of division, we might say that it is 5 because $5 \cdot 0 = 0$. We might also say that it is -8 because $-8 \cdot 0 = 0$. In fact, $0/0$ could be any number at all. So, division by 0 does not make sense. Division by 0 is not defined and is not possible.

EXAMPLES Divide, if possible.

4. $\dfrac{7}{0}$ Not defined.

5. $\dfrac{0}{7} = 0$ The quotient is 0 because $0 \cdot 7 = 0$.

(continued)

9 ▶ DIVIDE REAL NUMBERS (continued)

Two numbers whose product is 1 are called **reciprocals** (or **multiplicative inverses**). Every nonzero real number a has a **reciprocal** (or **multiplicative inverse**) $1/a$. The reciprocal of a positive number is positive. The reciprocal of a negative number is negative.

EXAMPLES   Find the reciprocal of each number.

6. $\dfrac{4}{5}$ The reciprocal is $\dfrac{5}{4}$, because $\dfrac{4}{5} \cdot \dfrac{5}{4} = 1$.

7. 8 The reciprocal is $\dfrac{1}{8}$, because $8 \cdot \dfrac{1}{8} = 1$.

8. $-\dfrac{2}{3}$ The reciprocal is $-\dfrac{3}{2}$, because $-\dfrac{2}{3} \cdot \left(-\dfrac{3}{2}\right) = 1$.

RECIPROCALS AND DIVISION

For any real numbers a and b, $b \neq 0$,

$$a \div b = \dfrac{a}{b} = a \cdot \dfrac{1}{b}.$$

(To divide, we can multiply by the reciprocal of the divisor.)

EXAMPLES   Divide by multiplying by the reciprocal of the divisor.

9. $\dfrac{1}{4} \div \dfrac{3}{5} = \dfrac{1}{4} \cdot \dfrac{5}{3} = \dfrac{5}{12}$ "Inverting" the divisor, $\dfrac{3}{5}$, and multiplying

10. $\dfrac{2}{3} \div \left(-\dfrac{4}{9}\right) = \dfrac{2}{3} \cdot \left(-\dfrac{9}{4}\right) = -\dfrac{18}{12}$, or $-\dfrac{3}{2}$

11. $-\dfrac{5}{7} \div 3 = -\dfrac{5}{7} \cdot \dfrac{1}{3} = -\dfrac{5}{21}$

For any numbers a and b, $b \neq 0$,

$$\dfrac{-a}{b} = \dfrac{a}{-b} = -\dfrac{a}{b} \quad \text{and} \quad \dfrac{-a}{-b} = \dfrac{a}{b}.$$

Do Exercises 1–14. ▶

EXERCISES

Divide, if possible.

1. $\dfrac{-8}{4}$

2. $\dfrac{-77}{-11}$

3. $\dfrac{5}{0}$

4. $\dfrac{0}{32}$

Find the reciprocal of each number.

5. $\dfrac{3}{4}$

6. $-\dfrac{7}{8}$

7. 25

8. 0.2

Divide.

9. $\dfrac{2}{7} \div \left(-\dfrac{11}{3}\right)$

10. $-\dfrac{10}{3} \div \left(-\dfrac{2}{15}\right)$

11. $-48 \div 0.4$

12. $\dfrac{8}{19} \div (-2)$

Find the opposite and the reciprocal.

13. $-\dfrac{3}{8}$

14. 6

Exponential notation is a shorthand device. For $3 \cdot 3 \cdot 3 \cdot 3$, we write 3^4. In the **exponential notation** 3^4, the number 3 is called the **base** and the number 4 is called the **exponent**. Exponential notation a^n, where n is an integer greater than 1, means

$$\underbrace{a \cdot a \cdot a \cdots a \cdot a.}$$

n factors

We read a^n as "a to the nth power," or simply "a to the nth." We can read a^2 as "a-squared" and a^3 as "a-cubed."

EXAMPLES Write exponential notation.

1. $7 \cdot 7 \cdot 7 = 7^3$ **2.** $xxxxx = x^5$ **3.** $\dfrac{2}{3} \cdot \dfrac{2}{3} \cdot \dfrac{2}{3} \cdot \dfrac{2}{3} = \left(\dfrac{2}{3}\right)^4$ ■

EXAMPLES Evaluate.

4. $5^3 = 5 \cdot 5 \cdot 5 = 125$ **5.** $\left(\dfrac{7}{8}\right)^2 = \dfrac{7}{8} \cdot \dfrac{7}{8} = \dfrac{49}{64}$

6. $-(10)^4 = -(10 \cdot 10 \cdot 10 \cdot 10) = -10{,}000$

7. $(-10)^4 = (-10)(-10)(-10)(-10) = 10{,}000$ ■

When an exponent is an integer greater than 1, it tells how many times the base occurs as a factor. What happens when the exponent is 1 or 0? Look for a pattern below. Think of dividing by 10 on the right.

$$10^3 = 10 \cdot 10 \cdot 10 = 1000$$
$$10^2 = 10 \cdot 10 = 100$$
$$10^1 = ?$$
$$10^0 = ?$$

In order for the pattern to continue, 10^1 would have to be 10 and 10^0 would have to be 1.

EXPONENTS OF 0 AND 1

For any number a, we agree that a^1 means a.

For any *nonzero* number a, we agree that a^0 means 1.

Mathematicians agree *not* to define 0^0.

EXAMPLES Rewrite without an exponent.

8. $4^1 = 4$ **9.** $(-97)^1 = -97$

10. $6^0 = 1$ **11.** $(-37.4)^0 = 1$ ■

EXERCISES

Write exponential notation.

1. $yyyyyy$

2. $8 \cdot 8 \cdot 8 \cdot 8$

3. $(3.8)(3.8)(3.8)(3.8)(3.8)$

4. $\left(-\dfrac{4}{5}\right)\left(-\dfrac{4}{5}\right)\left(-\dfrac{4}{5}\right)$

Evaluate.

5. $\left(\dfrac{3}{4}\right)^1$

6. 17^0

7. $(-2)^6$

8. $(-7)^3$

9. $\left(\dfrac{1}{3}\right)^3$

10. $-(3)^4$

11. $(0.1)^6$

12. $\left(-\dfrac{5}{7}\right)^2$

(continued)

10 ▸ EXPONENTIAL NOTATION (PART 1) (continued)

Look for a pattern below. Again, think of dividing by 10 on the right.

$$10^2 = 100$$
$$10^1 = 10$$
$$10^0 = 1$$
$$10^{-1} = ?$$
$$10^{-2} = ?$$

In order for the pattern to continue, 10^{-1} would have to be $\frac{1}{10}$ and 10^{-2} would have to be $\frac{1}{100}$.

NEGATIVE EXPONENTS
For any real number a that is nonzero and any integer n,

$$a^{-n} = \frac{1}{a^n}.$$

EXAMPLES Rewrite using a positive exponent. Evaluate, if possible.

12. $y^{-5} = \dfrac{1}{y^5}$

13. $\dfrac{1}{t^{-4}} = t^4$

14. $(-2)^{-3} = \dfrac{1}{(-2)^3} = \dfrac{1}{(-2)(-2)(-2)} = \dfrac{1}{-8} = -\dfrac{1}{8}$

15. $\left(\dfrac{2}{5}\right)^{-2} = \dfrac{1}{\left(\frac{2}{5}\right)^2} = \dfrac{1}{\frac{4}{25}} = 1 \cdot \dfrac{25}{4} = \dfrac{25}{4}$

················· Caution! ·················

A negative exponent does *not* necessarily indicate that an answer is negative! For example, 3^{-2} means $1/3^2$, or $1/9$, not -9.

EXAMPLES Rewrite using a negative exponent.

16. $\dfrac{1}{x^2} = x^{-2}$

17. $\dfrac{1}{(-7)^4} = (-7)^{-4}$

Do Exercises 1–20.
(Exercises 1–12 are on the preceding page.) ▶

EXERCISES

Rewrite using a positive exponent. Evaluate, if possible.

13. y^{-5}

14. $(-4)^{-3}$

15. $\left(\dfrac{3}{4}\right)^{-2}$

16. $\dfrac{1}{a^{-2}}$

17. $\left(\dfrac{1}{4}\right)^{-2}$

Rewrite using a negative exponent.

18. $\dfrac{1}{3^4}$

19. $\dfrac{1}{b^3}$

20. $\dfrac{1}{(-16)^2}$

MyLab Math
ANIMATION

RULES FOR ORDER OF OPERATIONS

1. Do all the calculations within grouping symbols, like parentheses, before operations outside.

2. Evaluate all exponential expressions.

3. Do all multiplications and divisions in order from left to right.

4. Do all additions and subtractions in order from left to right.

EXAMPLE 1 Simplify: $8 + 2 \cdot 5^3$.

$$8 + 2 \cdot 5^3 = 8 + 2 \cdot 125 \qquad \text{Evaluating the exponential expression}$$
$$= 8 + 250 \qquad \text{Doing the multiplication}$$
$$= 258 \qquad \text{Adding}$$

EXAMPLE 2 Simplify and compare: $(8 - 10)^2$ and $8^2 - 10^2$.

$$(8 - 10)^2 = (-2)^2 = 4;$$
$$8^2 - 10^2 = 64 - 100 = -36$$

We see that $(8 - 10)^2$ and $8^2 - 10^2$ are *not* the same.

EXAMPLE 3 Simplify: $3^4 + 62 \cdot 8 - 2(29 + 33 \cdot 4)$.

$$3^4 + 62 \cdot 8 - 2(29 + 33 \cdot 4)$$
$$= 3^4 + 62 \cdot 8 - 2(29 + 132) \qquad \text{Carrying out operations inside parentheses first; doing the multiplication}$$
$$= 3^4 + 62 \cdot 8 - 2(161) \qquad \text{Adding inside parentheses}$$
$$= 81 + 62 \cdot 8 - 2(161) \qquad \text{Evaluating the exponential expression}$$
$$= 81 + 496 - 2(161) \qquad \text{Doing the multiplications}$$
$$= 81 + 496 - 322 \qquad \text{in order from left to right}$$
$$= 577 - 322 \qquad \text{Doing all additions and}$$
$$= 255 \qquad \text{subtractions in order from left to right}$$

EXERCISES

Simplify.

1. $9[8 - 7(5 - 2)]$

2. $(9 - 12)^2$

3. $4 \div (8 - 10)^2 + 1$

4. $[64 \div (-4)] \div (-2)$

5. $9^2 - 12^2$

6. $[24 \div (-3)] \div \left(-\dfrac{1}{2}\right)$

7. $9 \div (-3) + 16 \div 8$

8. $20 + 4^3 \div (-8)$

9. $256 \div (-32) \div (-4)$

10. $9[(8 - 11) - 13]$

(continued)

11 ▶ ORDER OF OPERATIONS (continued)

When parentheses occur within parentheses, we can make them different shapes, such as [ ] (also called "brackets") and { } (usually called "braces"). Parentheses, brackets, and braces all have the same meaning. When parentheses occur within parentheses, **computations in the *innermost* ones are to be done first**.

EXAMPLE 4 Simplify: $5 - \{6 - [3 - (7 + 3)]\}$.

$$
\begin{aligned}
5 - \{6 - [3 - (7 + 3)]\} &= 5 - \{6 - [3 - 10]\} && \text{Adding } 7 + 3 \\
&= 5 - \{6 - [-7]\} && \text{Subtracting } 3 - 10 \\
&= 5 - 13 && \text{Subtracting } 6 - [-7] \\
&= -8
\end{aligned}
$$

EXAMPLE 5 Simplify: $\dfrac{12|7 - 9| + 8 \cdot 5}{3^2 + 2^3}$.

We do the calculations in the numerator and in the denominator separately, and then divide the results:

$$
\begin{aligned}
\frac{12|7 - 9| + 8 \cdot 5}{3^2 + 2^3} &= \frac{12|-2| + 8 \cdot 5}{9 + 8} \\
&= \frac{12(2) + 8 \cdot 5}{17} \\
&= \frac{24 + 40}{17} \\
&= \frac{64}{17}.
\end{aligned}
$$

Subtracting inside the absolute-value signs before taking the absolute value

Do Exercises 1–20.
(Exercises 1–10 are on the preceding page.) ▶

EXERCISES

Simplify.

11. $18 - 2 \cdot 3 - 9$

12. $(13 \cdot 2 - 8 \cdot 4)^2$

13. $12 - 4(5 - 1)$

14. $[2 \cdot (5 - 3)]^2$

15. $[5(8 - 6) + 12] - [24 - (8 - 4)]$

16. $-32 - 18 \div [4 - (-2)]$

17. $2^5 - [4 \cdot 3 - 2(1 + 4 \cdot 2)]$

18. $10 - 3\{8 \div [4 - (11 - 6)]\}$

19. $\dfrac{4|6 - 7| - 5 \cdot 4}{6 \cdot 7 - 8|4 - 1|}$

20. $\dfrac{(8 - 3)^2 + (7 - 10)^2}{3^2 - 2^3}$

12 ▶ TRANSLATE TO AN ALGEBRAIC EXPRESSION

MyLab Math

VIDEO

When a letter is used to represent various numbers, it is called a **variable**. Let $t =$ the number of hours that a passenger jet has been flying. Then t is a variable, because t changes as the flight continues. If a letter represents one particular number, it is called a **constant**. Let $d =$ the number of hours in a day. Then d is a constant.

An **algebraic expression** consists of variables, numbers, and operation signs, such as $+, -, \cdot, \div$. When an equals sign, $=$, is placed between two expressions, an **equation** is formed.

(continued)

To translate problems to equations, we need to know that certain words correspond to certain symbols, as shown in the following table.

Key Words

ADDITION	SUBTRACTION	MULTIPLICATION	DIVISION
add	subtract	multiply	divide
sum	difference	product	quotient
plus	minus	times	divided by
total	decreased by	twice	ratio
increased by	less than	of	per
more than			

Expressions like rs represent products and can also be written as $r \cdot s$, $r \times s$, $(r)(s)$, or $r(s)$. The multipliers r and s are also called **factors**.

A quotient $m \div 5$ can also be represented as $m/5$ or $\dfrac{m}{5}$.

EXAMPLE 1   Translate to an algebraic expression:

Eight less than some number.

We can use any variable we wish, such as x, y, t, m, n, and so on. Here we let t represent the number. If we knew the number to be 23, then the translation of "eight less than 23" would be $23 - 8$. If we knew the number to be 345, then the translation of "eight less than 345" would be $345 - 8$. Since we are using a variable for the number, the translation is

$t - 8$.   *Caution!* $8 - t$ would be incorrect.

EXAMPLES   Translate each of the following to an algebraic expression.

Phrase	*Algebraic Expression*
2. Half *of* a number	$\dfrac{1}{2}t$, or $\dfrac{t}{2}$
3. Five *more than* three *times* some number	$3p + 5$, or $5 + 3p$
4. The *difference* of two numbers	$x - y$
5. Seventy-six percent *of* some number	$76\%z$, or $0.76z$, or $\dfrac{76}{100}z$
6. Eight *less than twice* some number	$2x - 8$

Do Exercises 1–8. ▶

Do Exercises 1–8. ▶

EXERCISES

Translate each phrase to an algebraic expression.

1. x divided by 6

2. Seven more than four times some number

3. 25 less than t

4. One-third of a number

5. The sum of w and twice q

6. -18 multiplied by m

7. m subtracted from s

8. Nineteen percent of some number

Just-in-Time Review

EVALUATE ALGEBRAIC EXPRESSIONS

When we replace a variable with a number, we say that we are **substituting** for the variable. Carrying out the resulting calculation is called **evaluating the expression**. The result is called the **value** of the expression.

EXAMPLE 1 Evaluate $x - y$ when $x = 83$ and $y = 49$.

We substitute 83 for x and 49 for y and carry out the subtraction:

$$x - y = 83 - 49 = 34.$$

EXAMPLE 2 Evaluate a/b when $a = -63$ and $b = 7$.

We substitute -63 for a and 7 for b and carry out the division:

$$\frac{a}{b} = \frac{-63}{7} = -9.$$

EXAMPLE 3 Evaluate the expression $3xy + z$ when $x = 2$, $y = -5$, and $z = 7$.

We substitute and carry out the calculations according to the rules for order of operations:

$$3xy + z = 3(2)(-5) + 7 = -30 + 7 = -23.$$

EXAMPLE 4 Evaluate $5 + 2(a - 1)^2$ when $a = 4$.

$$\begin{aligned}
5 + 2(a - 1)^2 &= 5 + 2(4 - 1)^2 && \text{Substituting} \\
&= 5 + 2(3)^2 && \text{Working within} \\
& && \text{parentheses first} \\
&= 5 + 2(9) && \text{Simplifying } 3^2 \\
&= 5 + 18 && \text{Multiplying} \\
&= 23 && \text{Adding}
\end{aligned}$$

EXAMPLE 5 Evaluate $9 - x^3 + 6 \div 2y^2$ when $x = 2$ and $y = 5$.

$$\begin{aligned}
9 - x^3 + 6 \div 2y^2 &= 9 - 2^3 + 6 \div 2(5)^2 && \text{Substituting} \\
&= 9 - 8 + 6 \div 2 \cdot 25 && \text{Simplifying } 2^3 \\
& && \text{and } 5^2 \\
&= 9 - 8 + 3 \cdot 25 && \text{Dividing} \\
&= 9 - 8 + 75 && \text{Multiplying} \\
&= 1 + 75 && \text{Subtracting} \\
&= 76 && \text{Adding}
\end{aligned}$$

Do Exercises 1–8. ▶

EXERCISES

Evaluate.

1. $57y$, when $y = -8$

2. $\dfrac{x}{y}$, when $x = 30$ and $y = -6$

3. $\dfrac{5}{p + q}$, when $p = 20$ and $q = 30$

4. $\dfrac{18m}{n}$, when $m = 7$ and $n = 18$

5. $4x - y$, when $x = 3$ and $y = -2$

6. $2c \div 3b$, when $b = 4$ and $c = 6$

7. $25 - r^2 + s \div r^2$, when $r = 3$ and $s = 27$

8. $m + n(5 + n^2)$, when $m = 15$ and $n = 3$

EQUIVALENT FRACTION EXPRESSIONS

Two expressions that have the same value for all *allowable* replacements are called **equivalent expressions**.

THE IDENTITY PROPERTY OF 1

For any real number a, $a \cdot 1 = 1 \cdot a = a$.

(The number 1 is the **multiplicative identity**.)

We will often refer to the use of the identity property of 1 as "multiplying by 1."

EXAMPLE 1 Use multiplying by 1 to find an expression equivalent to $\frac{3}{5}$ with a denominator of $10x$.

Because $10x = 5 \cdot 2x$, we multiply by 1, using $2x/(2x)$ as a name for 1:

$$\frac{3}{5} = \frac{3}{5} \cdot 1 = \frac{3}{5} \cdot \frac{2x}{2x} = \frac{3 \cdot 2x}{5 \cdot 2x} = \frac{6x}{10x}.$$

Note that the expressions $3/5$ and $6x/(10x)$ are equivalent. They have the same value for any allowable replacement. Note too that 0 is not an allowable replacement in $6x/(10x)$, but for all nonzero real numbers, the expressions $3/5$ and $6x/(10x)$ have the same value.

In algebra, we consider an expression like $3/5$ to be a "simplified" form of $6x/(10x)$. To find such simplified expressions, we reverse the identity property of 1 in order to "remove a factor of 1."

EXAMPLE 2 Simplify: $\frac{7x}{9x}$.

$$\frac{7x}{9x} = \frac{7 \cdot x}{9 \cdot x} = \frac{7}{9} \cdot \frac{x}{x} = \frac{7}{9} \cdot 1 = \frac{7}{9}$$

EXAMPLE 3 Simplify: $-\frac{24y}{16y}$.

$$-\frac{24y}{16y} = -\frac{3 \cdot 8y}{2 \cdot 8y} = -\frac{3}{2} \cdot \frac{8y}{8y} = -\frac{3}{2} \cdot 1 = -\frac{3}{2}$$

Do Exercises 1–8. ▶

EXERCISES

Use multiplying by 1 to find an equivalent expression with the given denominator.

1. $\frac{7}{8}$; $8x$

2. $\frac{3}{10}$; $50y$

3. $\frac{1}{17}$; $51b$

4. $\frac{300}{49}$; $98w$

Simplify.

5. $\frac{25x}{15x}$

6. $\frac{36y}{18y}$

7. $-\frac{100a}{25a}$

8. $\frac{-625t}{15t}$

15 ▸ THE COMMUTATIVE LAWS AND THE ASSOCIATIVE LAWS

EXAMPLE 1 Evaluate $x + y$ and $y + x$ when $x = 5$ and $y = 8$.

$$x + y = 5 + 8 = 13; \qquad y + x = 8 + 5 = 13.$$

EXAMPLE 2 Evaluate xy and yx when $x = 4$ and $y = 3$.

$$xy = 4 \cdot 3 = 12; \qquad yx = 3 \cdot 4 = 12.$$

When only addition is involved, changing the order does not change the answer. Likewise, when only multiplication is involved, changing the order does not change the answer. The expressions $x + y$ and $y + x$ are equivalent, and the expressions xy and yx are equivalent.

THE COMMUTATIVE LAWS

Addition. For any numbers a and b,

$$a + b = b + a.$$

Multiplication. For any numbers a and b,

$$ab = ba.$$

EXAMPLE 3 Evaluate $a + (b + c)$ and $(a + b) + c$ when $a = 4$, $b = 8$, and $c = 5$.

$$a + (b + c) = 4 + (8 + 5) = 4 + 13 = 17;$$
$$(a + b) + c = (4 + 8) + 5 = 12 + 5 = 17$$

EXAMPLE 4 Evaluate $a \cdot (b \cdot c)$ and $(a \cdot b) \cdot c$ when $a = 7$, $b = 4$, and $c = 2$.

$$a \cdot (b \cdot c) = 7 \cdot (4 \cdot 2) = 7 \cdot 8 = 56;$$
$$(a \cdot b) \cdot c = (7 \cdot 4) \cdot 2 = 28 \cdot 2 = 56$$

When only addition is involved, changing the grouping does not change the answer. Likewise, when only multiplication is involved, changing the grouping does not change the answer. The expressions $a + (b + c) = (a + b) + c$ are equivalent, and the expressions $a \cdot (b \cdot c)$ and $(a \cdot b) \cdot c$ are equivalent.

THE ASSOCIATIVE LAWS

Addition. For any numbers a, b, and c,

$$a + (b + c) = (a + b) + c.$$

Multiplication. For any numbers a, b, and c,

$$a \cdot (b \cdot c) = (a \cdot b) \cdot c.$$

EXERCISES

Use a commutative law to find an equivalent expression.

1. $w + 3$

2. rt

3. $pq + 14$

Use an associative law to find an equivalent expression.

4. $m + (n + 2)$

5. $(7 \cdot x) \cdot y$

(continued)

EXAMPLE 5 Use the commutative laws and the associative laws to write at least three expressions equivalent to $(x + 8) + y$.

a) $(x + 8) + y = x + (8 + y)$

$= x + (y + 8)$

Using the associative law first and then the commutative law

b) $(x + 8) + y = y + (x + 8)$

$= y + (8 + x)$

Using the commutative law and then the commutative law again

c) $(x + 8) + y = (8 + x) + y$

$= 8 + (x + y)$

Using the commutative law first and then the associative law

Do Exercises 1–7.
(Exercises 1–5 are on the preceding page.) ▷

EXERCISES

Use the commutative laws and the associative laws to find three equivalent expressions.

6. $(a + b) + 8$

7. $9 \cdot (x \cdot y)$

---

16 THE DISTRIBUTIVE LAWS

MyLab Math

VIDEO

EXAMPLE 1 Evaluate $8(x + y)$ and $8x + 8y$ when $x = 4$ and $y = 5$.

$8(x + y) = 8(4 + 5)$ $8x + 8y = 8 \cdot 4 + 8 \cdot 5$

$= 8(9)$ $= 32 + 40$

$= 72;$ ⟵⟶ $= 72$

The expressions $8(x + y)$ and $8x + 8y$ are equivalent. ▪

THE DISTRIBUTIVE LAWS

The Distributive Law of Multiplication Over Addition

For any numbers a, b, and c,

$$a(b + c) = ab + ac, \quad \text{or} \quad (b + c)a = ba + ca.$$

The Distributive Law of Multiplication Over Subtraction

For any real numbers a, b, and c,

$$a(b - c) = ab - ac, \quad \text{or} \quad (b - c)a = ba - ca.$$

EXERCISES

Multiply.

1. $3(c + 1)$

2. $3(x - y)$

3. $-2(3c + 5d)$

4. $5x(y - z + w)$

5. $\dfrac{1}{2}h(a + b)$

6. $2\pi r(h + 1)$

(continued)

16 THE DISTRIBUTIVE LAWS (continued)

The distributive laws are the basis of multiplication in algebra as well as in arithmetic. In the following examples, note that we multiply each number or letter inside the parentheses by the factor outside.

EXAMPLES Multiply.

2. $4(x - 2) = 4 \cdot x - 4 \cdot 2 = 4x - 8$
3. $b(s - t + f) = bs - bt + bf$
4. $-3(y + 4) = -3 \cdot y + (-3) \cdot 4 = -3y - 12$
5. $-2x(y - 1) = -2x \cdot y - (-2x) \cdot 1 = -2xy + 2x$

The reverse of multiplying is called **factoring**. Factoring an expression involves factoring its *terms*. **Terms** of algebraic expressions are the parts separated by addition signs.

EXAMPLE 6 List the terms of $3x - 4y - 2z$.

We first find an equivalent expression that uses addition signs:

$3x - 4y - 2z = 3x + (-4y) + (-2z)$.   Using the property $a - b = a + (-b)$

Thus the terms are $3x$, $-4y$, and $-2z$.

To **factor** an expression is to find an equivalent expression that is a product. If $N = a \cdot b$, then a and b are **factors** of N.

EXAMPLES Factor.

7. $8x + 8y = 8(x + y)$ 8 and $x + y$ are factors.
8. $cx - cy = c(x - y)$ c and $x - y$ are factors.

Generally, we try to factor out the largest factor common to all the terms. In the following example, we might factor out 3, but there is a larger factor common to the terms, 9. So we factor out the 9.

EXAMPLE 9 Factor: $9x + 27y$.

$9x + 27y = 9 \cdot x + 9 \cdot (3y) = 9(x + 3y)$

We often must supply a factor of 1 when factoring out a common factor, as in the next example, which is a formula involving simple interest.

EXAMPLE 10 Factor: $P + Prt$.

$P + Prt = P \cdot 1 + P \cdot rt$ Writing P as a product of P and 1
$= P(1 + rt)$ Using a distributive law

Do Exercises 1–14.
(Exercises 1–6 are on the preceding page.) ▶

EXERCISES

List the terms.

7. $4a - 5b + 6$

8. $2x - y - 1$

Factor.

9. $9a + 9b$

10. $7x - 21$

11. $ab + a$

12. $18a - 24b - 48$

13. $8m + 4n - 24$

14. $xy - xz + xw$

If two terms have the same letter, or letters, we say that they are **like terms**, or **similar terms**. (If powers, or exponents, are involved, then like terms must have the same letters raised to the same powers.)

    If two terms have no letters at all but are just numbers, they are also similar terms. We can simplify by **collecting**, or **combining**, **like terms**, using a distributive law.

EXAMPLES   Collect like terms.

1. $3x + 5x = (3 + 5)x = 8x$ Factoring out the x using a distributive law

2. $x - 3x = 1 \cdot x - 3 \cdot x = (1 - 3)x = -2x$ $x = 1 \cdot x$

3. $2x + 3y - 5x - 2y$

$\quad = 2x - 5x + 3y - 2y$ Using a commutative law

$\quad = (2 - 5)x + (3 - 2)y$ Using a distributive law

$\quad = -3x + y$ Simplifying

4. $3x + 2x + 5 + 7 = (3 + 2)x + (5 + 7) = 5x + 12$

5. $4.2x - 6.7y - 5.8x + 23y = (4.2 - 5.8)x + (-6.7 + 23)y$

$\qquad\qquad\qquad\qquad\qquad\quad = -1.6x + 16.3y$

6. $-\dfrac{1}{4}a + \dfrac{1}{2}b - \dfrac{3}{5}a - \dfrac{2}{5}b = \left(-\dfrac{1}{4} - \dfrac{3}{5}\right)a + \left(\dfrac{1}{2} - \dfrac{2}{5}\right)b$

$\qquad\qquad\qquad\qquad\qquad = \left(-\dfrac{5}{20} - \dfrac{12}{20}\right)a + \left(\dfrac{5}{10} - \dfrac{4}{10}\right)b$

$\qquad\qquad\qquad\qquad\qquad = -\dfrac{17}{20}a + \dfrac{1}{10}b$

Do Exercises 1–10. ▶

EXERCISES

Collect like terms.

1. $7x + 5x$

2. $8b - 11b$

3. $14y + y$

4. $12a - a$

5. $t - 9t$

6. $5x - 3x + 8x$

7. $3c + 8d - 7c + 4d$

8. $4x - 7 + 18x + 25$

9. $1.3x + 1.4y - 0.11x - 0.47y$

10. $\dfrac{2}{3}a + \dfrac{5}{6}b - 27 - \dfrac{4}{5}a - \dfrac{7}{6}b$

18 ▶ **REMOVING PARENTHESES AND COLLECTING LIKE TERMS**

MyLab Math

VIDEO

THE PROPERTY OF −1

For any number a,

$$-1 \cdot a = -a.$$

(Negative 1 times a is the opposite of a. In other words, changing the sign is the same as multiplying by −1.)

EXAMPLES   Find an equivalent expression without parentheses.

1. $-(3x) = -1(3x)$ Replacing − with −1 using the property of −1

$\quad\quad = (-1 \cdot 3)x$ Using an associative law

$\quad\quad = -3x$ Multiplying

2. $-(-9y) = -1(-9y) = [-1(-9)]y = 9y$

3. $-(4 + x) = -1(4 + x)$ Replacing − with −1

$\quad\quad = -1 \cdot 4 + (-1) \cdot x$ Multiplying using the distributive law

$\quad\quad = -4 + (-x)$ Replacing $-1 \cdot x$ with $-x$

$\quad\quad = -4 - x$ Adding an opposite is the same as subtracting.

4. $-(a - b) = -1(a - b) = -1 \cdot a - (-1) \cdot b$

$\quad\quad = -a + [-(-1)b] = -a + b, \text{ or } b - a$

For any real numbers a and b,

$$-(a - b) = b - a.$$

(The opposite of $a - b$ is $b - a$.)

We can find an equivalent expression for an opposite by multiplying every term by −1. We could also say that we change the sign of every term inside the parentheses.

EXAMPLE 5   Find an equivalent expression without parentheses:

$$-\left(-9t + 7z - \tfrac{1}{4}w\right).$$

We have

$$-\left(-9t + 7z - \tfrac{1}{4}w\right) = 9t - 7z + \tfrac{1}{4}w. \quad \text{Changing the sign of every term}$$

EXERCISES

Find an equivalent expression without parentheses.

1. $-(-2c)$

2. $-(b + 4)$

3. $-(x - 8)$

4. $-(t - y)$

5. $-(r + s + t)$

6. $-(8x - 6y + 13)$

(continued)

Some expressions contain parentheses preceded by subtraction signs. These parentheses can be removed by changing the sign of *every* term inside.

EXAMPLES   Remove parentheses and simplify.

6. $6x - (4x + 2) = 6x - 4x - 2$ Changing the sign of every term inside parentheses

$\qquad\qquad\qquad = 2x - 2$ Collecting like terms

7. $3y - 4 - (9y - 7) = 3y - 4 - 9y + 7$

$\qquad\qquad\qquad\qquad = -6y + 3, \text{ or } 3 - 6y$

8. $3y + (3x - 8) - (5 - 12y) = 3y + 3x - 8 - 5 + 12y$

$\qquad\qquad\qquad\qquad\qquad = 15y + 3x - 13$

9. $x - 3(x + y) = x - 3x - 3y$ Removing parentheses by multiplying $x + y$ by -3

$\qquad\qquad\quad = -2x - 3y$ Collecting like terms

·········· **Caution!** ··········

A common error is to forget to change this sign. *Remember*: When multiplying by a negative number, change the sign of *every* term inside the parentheses.

10. $3y - 2(4y - 5) = 3y - 8y + 10 = -5y + 10$

When expressions with grouping symbols contain variables, we still work from the inside out when simplifying, using the rules for order of operations.

EXAMPLE 11   Simplify: $[2(x + 7) - 4^2] - (2 - x)$.

$[2(x + 7) - 4^2] - (2 - x)$

$= [2x + 14 - 4^2] - (2 - x)$ Multiplying to remove the innermost grouping symbols using a distributive law

$= [2x + 14 - 16] - (2 - x)$ Evaluating the exponential expression

$= [2x - 2] - (2 - x)$ Collecting like terms inside the brackets

$= 2x - 2 - 2 + x$ Multiplying by -1 to remove the parentheses

$= 3x - 4$ Collecting like terms

Do Exercises 1–12.
(Exercises 1–6 are on the preceding page.) ▷

EXERCISES

Simplify.

7. $4m - (3m - 1)$

8. $7a - [9 - 3(5a - 2)]$

9. $-2(x + 3) - 5(x - 4)$

10. $8x - (-3y + 7) + (9x - 11)$

11. $[7(x + 5) - 19] - [4(x - 6) + 10]$

12. $5\{-2 + 3[4 - 2(3 + 5)]\}$

19 ▶ EXPONENTIAL NOTATION (PART 2)

We often need to find ways to determine *equivalent exponential expressions*. We do this with several rules or properties regarding exponents.

RULES FOR EXPONENTS

For any real numbers a and b and any integers m and n:

Product Rule: $\qquad a^m \cdot a^n = a^{m+n}$

Quotient Rule: $\qquad \dfrac{a^m}{a^n} = a^{m-n}, a \neq 0$

Power Rule: $\qquad (a^m)^n = a^{mn}$

Raising a product to a power: $\quad (ab)^n = a^n b^n$

Raising a quotient to a power: $\quad \left(\dfrac{a}{b}\right)^n = \dfrac{a^n}{b^n}, b \neq 0;$

$$\left(\dfrac{a}{b}\right)^{-n} = \left(\dfrac{b}{a}\right)^n, b \neq 0, a \neq 0$$

EXAMPLES Simplify.

1. $x^4 \cdot x^3 = x^{4+3} = x^7 \qquad$ Adding exponents using the product rule

2. $\dfrac{5^7}{5^3} = 5^{7-3} = 5^4 \qquad$ Subtracting exponents using the quotient rule

3. $\dfrac{5^7}{5^{-3}} = 5^{7-(-3)} = 5^{7+3} = 5^{10} \qquad$ Subtracting exponents (adding an opposite)

4. $4^5 \cdot 4^{-3} = 4^{5+(-3)} = 4^2 = 16 \qquad$ Adding exponents

5. $\dfrac{9^{-2}}{9^5} = 9^{-2-5} = 9^{-7} = \dfrac{1}{9^7} \leftarrow$

> Note that we give answers using positive exponents. In some situations, this may not be appropriate, but we do so here.

6. $(8x^n)(6x^{2n}) = 8 \cdot 6 \cdot x^n \cdot x^{2n} = 48 \cdot x^{n+2n} = 48x^{3n}$

7. $\dfrac{16x^4y^7}{-8x^3y^9} = \dfrac{16}{-8} \cdot \dfrac{x^4}{x^3} \cdot \dfrac{y^7}{y^9} = -2x^{4-3}y^{7-9} = -2x^1y^{-2} = -\dfrac{2x}{y^2}$

EXERCISES

Simplify.

1. $8^{-2} \cdot 8^{-4}$

2. $\dfrac{a^3}{a^{-2}}$

3. $\dfrac{10^{-3}}{10^6}$

4. $b^2 \cdot b^{-5}$

5. $\dfrac{-24x^6y^7}{18x^{-3}y^9}$

6. $(14m^2n^3)(-2m^3n^2)$

7. $(-3t^{-4a})(-5t^{-a})$

8. $\dfrac{-18x^{-2}y^3}{-12x^{-5}y^5}$

(continued)

8. $\dfrac{14x^5y^{-3}}{4x^9y^{-5}} = \dfrac{14}{4} \cdot \dfrac{x^5}{x^9} \cdot \dfrac{y^{-3}}{y^{-5}} = \dfrac{7}{2}x^{5-9}y^{-3-(-5)}$

$\qquad = \dfrac{7}{2}x^{-4}y^2 = \dfrac{7y^2}{2x^4}$

9. $(8x^4y^{-2})(-3x^{-3}y) = 8 \cdot (-3) \cdot x^4 \cdot x^{-3} \cdot y^{-2} \cdot y^1$

$\qquad = -24x^{4+(-3)}y^{-2+1}$

$\qquad = -24xy^{-1} = -\dfrac{24x}{y}$

When an expression inside parentheses is raised to a power, the inside expression is the base. Thus, for $(5^2)^4$, we are raising 5^2 to the fourth power:

$$(5^2)^4 = (5^2)(5^2)(5^2)(5^2) = 5^8.$$

EXAMPLES Simplify.

10. $(x^5)^7 = x^{5 \cdot 7} = x^{35}$ \qquad Multiplying exponents

11. $\left(\dfrac{3^{-4}}{5^2}\right)^6 = \dfrac{(3^{-4})^6}{(5^2)^6} = \dfrac{3^{-24}}{5^{12}} = \dfrac{1}{3^{24} \cdot 5^{12}}$

12. $(x^4)^{-2t} = x^{4(-2t)} = x^{-8t} = \dfrac{1}{x^{8t}}$

13. $\left(\dfrac{x^2}{y^{-3}}\right)^{-5} = \dfrac{x^{2 \cdot (-5)}}{y^{-3 \cdot (-5)}} = \dfrac{x^{-10}}{y^{15}} = \dfrac{1}{x^{10}y^{15}}$

14. $\left(\dfrac{-2a^{-5}}{a^{-3}b}\right)^4 = \dfrac{(-2a^{-5})^4}{(a^{-3}b)^4} = \dfrac{(-2)^4(a^{-5})^4}{(a^{-3})^4b^4}$

$\qquad = \dfrac{16a^{-20}}{a^{-12}b^4} = \dfrac{16a^{-20-(-12)}}{b^4}$

$\qquad = \dfrac{16a^{-8}}{b^4} = \dfrac{16}{a^8b^4}$

15. $(4r^3y^{-2}w)^{-2} = 4^{-2}(r^3)^{-2}(y^{-2})^{-2}w^{-2}$

$\qquad = \dfrac{1}{4^2}r^{-6}y^4w^{-2}$

$\qquad = \dfrac{y^4}{16r^6w^2}$

Do Exercises 1–16.
(Exercises 1–8 are on the preceding page.) ▶

EXERCISES

Simplify.

9. $(r^9)^5$

10. $\left(\dfrac{4^{-3}}{3^4}\right)^3$

11. $(5a^2b^2)^3$

12. $\left(\dfrac{a^4}{c^7}\right)^2$

13. $(4^3)^2$

14. $\left(\dfrac{2x^3y^{-2}}{3y^{-3}}\right)^3$

15. $(8^x)^{4y}$

16. $\left(\dfrac{-4x^4y^{-2}}{5x^{-1}y^4}\right)^{-4}$

20 ▶ SCIENTIFIC NOTATION

Scientific notation is useful when representing very large numbers and very small numbers and when estimating. The following are examples of scientific notation:

- Brazil spent $9,700,000,000 to stage the Summer 2016 Olympic Games.
  Data: Olympic Organizing Committee, Public Olympic Authority of Brazil, Federal Government of Brazil

  $\$9,700,000,000 = \$9.7 \times 10^9$

- The diameter of a helium atom is about 0.000000022 cm.

  $0.000000022 \text{ cm} = 2.2 \times 10^{-8} \text{ cm}$

Scientific notation for a number is an expression of the type

$$M \times 10^n,$$

where n is an integer, M is greater than or equal to 1 and less than 10 $(1 \le M < 10)$, and M is expressed in decimal notation. 10^n is also considered to be scientific notation when $M = 1$.

You should try to make conversions to scientific notation mentally as much as possible. A positive exponent in scientific notation indicates a large number (greater than or equal to 10) and a negative exponent indicates a small number (between 0 and 1).

EXAMPLES Convert mentally to scientific notation.

1. Light travels 9,460,000,000,000 km in one year.

   $9,460,000,000,000 = 9.46 \times 10^{12}$ 9.460,000,000,000.

   12 places

   Large number, so the exponent is positive.

2. The mass of a grain of sand is 0.0648 g (grams).

   $0.0648 = 6.48 \times 10^{-2}$ 0.06.48

   2 places

   Small number, so the exponent is negative.

EXERCISES

Convert each number to scientific notation.

1. 47,000,000,000

2. 0.000000016

3. 0.000000263

4. 2,600,000,000,000

5. *Insect-Eating Lizard.* A gecko is an insect-eating lizard. Its feet will adhere to virtually any surface because they contain millions of miniscule hairs, or setae, that are 200 billionths of a meter wide. Write 200 billionths in scientific notation.
   Data: *The Proceedings of the National Academy of Sciences*, Dr. Kellar Autumn and Wendy Hansen of Lewis and Clark College, Portland, Oregon

(continued)